Wisdom
of the
Quran

Wisdom
of the
Quran

Does science escape what Quran teaches us?

Addressed to Youth of Our Millennium

وَأَوْحَىٰ رَبُّكَ إِلَى ٱلنَّحْلِ أَنِ ٱتَّخِذِى مِنَ ٱلْجِبَالِ بُيُوتًا وَمِنَ ٱلشَّجَرِ وَمِمَّا يَعْرِشُونَ
Quran 16:68

MAHAMADOU S TOURE

WRITER AND STUDENT OF ISLAMIC WISDOM

BabouConnect Publishing

Published by: **BabouConnect Publishing**

ISBN (Paperback) : **9798993907277**

For permissions, inquiries, or bulk orders, contact:
sabilfalah@gmail.com

This is a work of nonfiction. The events and figures depicted are based on historical facts and author research. Any errors or omissions are unintentional.

Edition Year: 2025

This book was originally written by the author in:

Arabic (2014):

حكمة آيات الله

للشباب في القرن الحالي

French (2016):
La Sagesse du Coran

Contents

Preface

❋ ❋ ❋

Mahamadou S. TOURE was born in Mali, the son of **El Hadj Saad Oumar TOURE**, a pioneering reformer of Islamic education in West Africa. From an early age, he was immersed in Islamic teachings, beginning his studies at home with his mother, a Quranic teacher, and later at the **Sebile El Falah Institute**, founded by his father in Ségou, Mali.

Fluent in **Arabic and French**, Mahamadou earned high school diplomas in both languages, allowing him to first write this book in Arabic before translating it into French. His commitment to knowledge led him to **Ohio State University**, where he studied **public affairs** and specialized in **NGO management**.

Dedicated to promoting Islam's **values of peace, tolerance, and knowledge,** he emphasizes the harmony between **Islamic teachings and modern science.** His work serves as a guiding light, particularly for young Muslims, in understanding the **timeless relevance of the Quran.**

Foreword

The Most Exalted said:

"We will show them Our signs in the horizons and within themselves until it becomes clear to them that this is the Truth." (Quran, Chapter 41, Verse 53)

All praise is due to Allah, the Most High—the One and Only God, without partner, and the Sole Creator of all existence. He is the Source of boundless happiness in this world, granting His divine excellence to whomever He wills, believers and nonbelievers alike. Allah is the All-Knowing, who taught by the pen and illuminated the righteous path with His sacred knowledge. Peace and blessings be upon the Prophet Muhammad, the truthful one who faithfully conveyed Allah's wisdom and the teachings of Islam to all of humanity. Through his eloquence and steadfastness, the light of divine guidance spread across generations, inspiring hearts and minds alike.

This book is an earnest endeavor to serve the religion of Allah and uphold the teachings of His noble Prophet (peace and blessings be upon him). It aims to remind Muslims of the profound lessons found in the Holy Quran, encompassing Islamic laws, the story of humankind from Adam, the lives of the prophets (peace be upon them) and their miracles, and above all, the wisdom of Allah Most High as demonstrated in His creation, including the universe.

This divine wisdom reveals the unmatched power and knowledge of the Creator and encourages humankind to engage in scientific exploration. Such endeavors must, however, rest on a foundation of belief in Allah the Most High and His divine attributes.

The Quran emphasizes:

"Those who remember Allah while standing, sitting, and lying on their sides, and reflect on the creation of the heavens and the earth, say: 'Our Lord, You have not created this in vain. Glory be to You! Protect us from the torment of the Fire." Quran, 3:191

This book seeks to demonstrate that science does not escape the teachings of the Quran, revealed 14 centuries ago. It poses an essential question: Can a Muslim afford to disregard science? Furthermore, it aims to inspire the youth of our millennium to recognize the importance of science in Islam and to appreciate Allah's signs, many of which are realized through the efforts of modern scientists.

Entitled *"Wisdom of the Quran: Does Science Escape What the Quran Teaches Us?"* this book is firmly rooted in authentic Quranic verses. Written in accessible language, it strives to ensure readers from all walks of life can fully grasp its message.

With humility, this work extends an open invitation to the esteemed scholars of Islam to review and correct any inaccuracies or inconsistencies within its pages. Indeed, all matters ultimately return to Allah the Most High, the Sovereign and Supreme Authority, and to His final and beloved messenger, Muhammad (peace and blessings be upon him), who illuminated the way for humanity to embrace the truth of Islam.

CHAPTER I:

❖ ❖ ❖

STIMULATION

Does Science

Escape What the Quran Teaches Us?

Science, through its perpetual development over the centuries, has led us to numerous discoveries and conclusions about phenomena occurring in the heavens. Yet, the Holy Book, the Quran, in clear and direct statements, has drawn our attention to countless predictions, pieces of information, and initiatives for exploring the heavens, including those pertaining to most aspects of the universe on which science has always relied. This chapter invites readers to engage in a reflective debate: Does science truly escape what the Quran teaches us?

Dear readers, let us distinguish the originality and veracity of the Quran by examining the following categories side by side:

- **Human Science (recent centuries):** Exploitations, predictions, and discoveries

- **The Quran (7th century):** Information, predictions, and initiatives for exploration

---

1. Layers of the Heavens and Celestial Motion

Recent Centuries: When did science conclude that the heavens consist of layers of superimposed smoke? Modern observations by institutions like NASA affirm that the sun is a massive star, journeying along a predetermined path toward a specific destination within the solar system—a system intricately governing life on Earth. Such discoveries prompt profound reflection: how could truths uncovered by modern science align so precisely with descriptions revealed in the Quran.

7th Century Quranic Revelation:

a. *"And We have built above you seven strong [heavens] and placed therein a blazing lamp (the sun)."*

(Quran, 78:12-13)

b. *"And the sun runs on to its resting place. That is the determination of the All-Mighty, the All-Knowing."*

(Quran, 36:38)

---

2. Earth: The Unique Planet of Life

Recent Centuries: Science has consistently maintained that Earth is the only known planet capable of supporting life. Yet, the Quran articulated this singularity centuries before such discoveries were made.

7th Century Quranic Revelation:

"He it is Who made the earth subservient to you; so traverse its vast regions and eat of His provision. To Him will be the resurrection."

(Quran, 67:15)

3. The Origin and Role of Iron

Recent Centuries: Modern astrophysicists, such as Carl Sagan, have confirmed that iron, an essential element for human life, originated from the intense supernova explosions of dying stars. This cosmic iron descended from the heavens and was eventually integrated into Earth, demonstrating the intricate design of the universe. Scholars like Dr. Maurice Bucaille have pointed out the remarkable alignment of this discovery with Quranic revelations from the 7th century.

7th Century Quranic Revelation:

"And We sent down iron, wherein is great military might and benefits for the people, so that Allah may make evident those who support Him and His messengers unseen. Indeed, Allah is Powerful and Exalted in Might."

(Quran, 57:25)

4. The Upright Stature and Unique Creation of Humans

Recent Centuries: Scientists recognize that humans are the most upright, physically perfect mammals living in the safest environments. The Quran affirmed this unique status long before such observations:

7th Century Quranic Revelation:

> a. *"We have certainly created man in the best of stature."*

(Quran, 95:4)

> b. *"And Allah has made for you from your homes a place of rest and made for you tents from the hides of animals, which you find light when you travel and when you camp."* (Quran, 16:80)

---

5. The Mystery of Human Creation

Recent Centuries: The detailed study of human embryonic development began with the invention of the microscope in the 16th century. Since then, science has uncovered significant information about the transition from fetus to child. Yet, the Quran described these stages with astonishing clarity:

7th Century Quranic Revelation:

"We created man from an extract of clay. Then We made him a drop of fluid in a secure resting place. Then We created the fluid into a clinging clot, and We created the clot into a lump, and We created the lump into bones, and We clothed the bones with flesh. Then We developed him into another creation. So blessed is Allah, the Best of Creators."

(Quran, 23:12-14)

Even today, science remains unable to determine the exact moment during embryonic development when the soul enters the body. This limitation underscores the boundaries of human knowledge and the vastness of Allah's omnipotence.

Wisdom: In fact, there must be an invisible hand somewhere in nature—a hand endowed with supreme intelligence directing all of this. Each person may glorify this entity with a name, in any language. However, it would be more logical, coherent, and straightforward to recognize that this Supreme Knower can only be God, Lord of the world.

"And say, 'The truth is from your Lord, so whoever wills—let him believe; and whoever wills—let him disbelieve." (Quran, 18:29)

Peace upon Prophet Muhammad. It's fair to say that science can't escape from what the Quran teaches us!

Guidelines on the Importance of Science in Islam for the Youth of Our Millennium

Oh Muslim youth of this era of modern science, the primary objective of this book is to draw your attention to the immense value of science and its discoveries, which are integral to both worldly advancement and spiritual growth. For instance, consider the convenience of reading the Quran on phones or the ability to travel by air to visit the Mosque of Abraham (peace be upon him).

Let us praise Allah, the First Scientist, who has bestowed upon humanity a portion of His divine wisdom. Just as He empowered prophets (peace be upon them) with miracles, He has endowed us with the ability to innovate and explore.

While we must honor the scientists for their contributions, our ultimate gratitude belongs to Allah, the Creator of the universe, science, and the scientists themselves. The Quran beautifully states:

"It is Allah who created the heavens and the earth and sent down rain from the sky and produced thereby some fruits as provision for you and subjected for you the ships to sail through the sea by His command and subjected for you the rivers."

"And He subjected for you the sun and the moon, continuous [in orbit], and subjected for you the night and the day."

"And He gave you from all you asked of Him. And if you should count the favor of Allah, you could not enumerate them. Indeed, mankind is [generally] most unjust and ungrateful."

(Quran 14:32-34)

Wisdom: Science, therefore, becomes a pathway for humanity to worship Allah through reflection, gratitude, and obedience. It reminds us that everything we achieve originates from Allah, the Lord of the Worlds. The Quran consistently calls upon us to reflect on our existence and to venture into the heavens and earth in pursuit of understanding His creation.

1. The Mention of Modern Scientists in the Quran

Critical reflection on the heavens and nature—the signs of Allah—reveals that human science is but a fraction of the knowledge of Allah, the Supreme Scientist. The Quran affirms:

"We will show them Our signs in the universe and within themselves until it becomes clear to them that this [Quran] is the Truth. Is it not sufficient that your Lord is a witness over all things?"

(Quran 41:53)

"Indeed, in the creation of the heavens and the earth, and the alternation of the night and the

day, there are signs for those of understanding."
(Quran 3:190)

Modern scientists, inspired by this Quranic call to reflection, have ventured into space, revealing truths that align with divine wisdom. Their explorations demonstrate the majesty of Allah's creation, reminding us of the intricate design of the universe and our responsibility as stewards of this knowledge.

---

2. Scientists' Exploration of Space

"And the moon – We have set for it phases, until it returns

[appearing] like an old date stalk."

(Quran 36:39)

The Quran describes the moon's phases with remarkable accuracy, underscoring that such insights are not easily observable without scientific exploration. Additional verses provide further insight:

(a) "It is He who created seven heavens in layers." (Quran 67:3)

(b) "And by the moon when it becomes full, You will surely pass from one state to another. So what is [the matter] with them [that] they do not believe?"

(Quran 84:18-20)

These verses anticipate humanity's eventual ascension to the heavens—a prediction fulfilled through centuries of scientific discovery. From the understanding of the solar system to lunar exploration, these endeavors affirm the Quran's truth.

Yet, even as we uncover these mysteries, the Quran reminds us of the ultimate reality:

"Wherever you may be, death will overtake you, even if you should be within towers of lofty construction."

(Quran 4:78)

Wisdom: Allah, the First Scientist, graciously shares a fraction of His knowledge, enabling us to explore His creation. Science mirrors the divine system of accountability, as the universe and its intricacies ultimately point back to Allah's unmatched wisdom and authority.

---

3. Belief in Divine Revelation and Judgment Day Personal Reflection:

As a young Muslim, I recently renewed my belief in the divine revelation of the Quran through a simple experience with modern technology. While speaking with my mother across the Atlantic Ocean, I could instantly discern her mood—whether joyful or distressed—through her tone. This sensitivity, embedded in a small device, reflects the intelligence Allah has granted to humanity.

"Nor does he speak from [his own] inclination. It is not but a revelation revealed, taught to him by one intense in strength (Angel Gabriel)."

(Quran 53:3–5)

This alignment between modern scientific tools and Quranic truths reaffirms the divine origin of the Quran and reinforces trust in the Prophet Muhammad (peace be upon him), Allah's final messenger.

Judgment Day:

In today's world, governments assign unique identification numbers to individuals, allowing for comprehensive records of their actions. This human system mirrors the divine system of accountability revealed in the Quran:

God The Most High says:

"And every person's deeds We have fastened to his neck, and on the Day of Resurrection We will bring forth for him a record which he will find spread open. Read your record. Sufficient is yourself against you this Day as an accountant."

(Quran 17:13–14)

Belief in divine revelation and Judgment Day deepens our understanding of Allah's signs and emphasizes our responsibility as stewards of knowledge.

Why Are the

Youth My Primary Focus?

Simply because youth is a time of sharp intellect and strong inclination toward discovery—especially today, amidst diverse beliefs and rapid scientific progress.

Oh Muslim youth, embrace science as a means to strengthen your faith, reflect on Allah's signs, and contribute to humanity. Islam is, without a doubt, the one true religion of Allah, perfected through Prophet Muhammad (peace be upon him).

These words express my gratitude to Allah and my passion for contributing to the expansion of Islam. Yet, no Muslim—past, present, or future—could ever surpass the Prophet Muhammad (peace be upon him) in his tireless efforts to spread Allah's message and bring humanity toward the light of truth.

Finally, I dedicate this chapter to my dear father and teacher, El Hadj Saad Oumar Touré. May Allah grant him the highest place in Paradise.

CHAPTER II:

THE WISDOM OF GOD

Allah the Most-High, the Supreme Knower, possesses sacred knowledge and is the Genius Creator of all. His creation is flawless, and His wisdom is evident in every aspect of the universe. Through divine knowledge, miracles, and His boundless favor, Allah manifests His unmatched power and mercy.

The Quran declares:

"And We did not create the heavens and the earth

and everything between them except with truth and for a specific purpose." (Quran, 15:85)

This verse emphasizes that everything Allah created is purposeful, rooted in absolute truth, and reflects His divine wisdom.

I.   Divine Knowledge: Creation and   Miracles

A. Creation

Allah the Most High created the universe, vast and intricate, with no equal. He alone possesses complete knowledge of all things and has granted humanity the ability to learn and explore. However, our knowledge remains limited, serving as both a blessing and a test of gratitude. Driven by curiosity, humans have observed the heavens, stars, winds, and diverse forms of life, marveling at the extraordinary balance and order. These observations lead us to the inescapable truth: there must be a Supreme Creator—Allah, the Exalted and Omnipotent.

The Quran highlights this precision:

"He who created seven heavens in layers. You do not see any flaw in the creation of the Most Merciful. So return your vision; do you see any breaks? Then return your vision twice again— your vision will return to you humbled while it is fatigued."

(Quran, 67:3–4)

Even the symmetry of creation, the movement of celestial bodies, and the pairing of creatures point to Allah's design:

"So Abraham said, 'Indeed, Allah brings up the sun from the east, so bring it up from the west.' Thus the disbeliever was confounded. And Allah does not guide the wrongdoing people."

(Quran, 2:258)

The Wisdom of Human Endeavors:

Despite humanity's advancements in replicating aspects of Allah's creation—such as replacing sunlight with artificial light or utilizing iron in medicine and construction—these accomplishments serve as a testament to Allah's mercy. However, human achievements are limited by mortality and imperfection, further underscoring the greatness of Allah, the First and Supreme Scientist.

The Quran urges reflection:

"Those who remember Allah while standing, sitting, or lying down, and reflect on the creation of the heavens and the earth, say: 'Our Lord, You did not create this aimlessly; glory be to You! Protect us from the punishment of the Fire.' " *(Quran, 3:191)*

By contemplating these signs, we affirm that science, derived from divine wisdom, is a means to deepen our faith in Allah and understand His creation more profoundly.

B. Miracles and the Prophets

The Religion of Allah

Allah the Most High brought humanity from ignorance to the light of His religion by sending prophets, each endowed with unique miracles.

These miracles, granted by Allah's permission, served as evidence of His power and a call to worship Him alone.

The Quran affirms:

"And We did not send any messenger except speaking in the language of his people to state clearly for them."

(Quran, 14:4)

The Wisdom: The succession of prophets—from Adam to Muhammad (peace be upon them)—illustrates the consistency of Allah's religion. Though the names of these faiths varied across cultures and languages, the message remained the same: submission to Allah, the Creator.

Miracles of the Prophets:

1. **Abraham (peace be upon him):**
   Abraham survived being cast into a great fire by his enemies—a miracle that demonstrated Allah's protection. He also constructed the Kaaba and prayed at Mecca, leaving a lasting legacy for humanity.

"And Allah took Abraham as an intimate friend." (Quran, 4:125)

2. **Jesus (peace be upon him):** Jesus performed many miracles, including curing the blind, healing lepers, and even resurrecting the dead—all by Allah's permission. The Quran states:

"And I cure the blind and the leper, and I give life to the dead—by the permission of Allah."

(Quran, 3:49)

3. **Muhammad (peace be upon him):**

The greatest miracle of the Prophet Muhammad (peace be upon him) is the Quran, a divine revelation that continues to guide humanity. Other notable miracles include the splitting of the moon and the Night Journey (Isra and Miraj), which demonstrated his connection to Allah's divine will.

"Blessed is He who sent down the Criterion (the Quran) to His servant that he may be a warner to the worlds."

(Quran, 25:1)

C. Wisdom of Prophet Muhammad (peace be upon him):

From his noble character and teachings to the miracles performed with Allah's permission, every aspect of his prophethood serves as an illustration of the wisdom of the Quran. By following his example, both Muslims and non-Muslims can achieve a balance between faith, science, and daily living, fulfilling their role as stewards of Allah's creation.

1. Character and Habits:

The Prophet Muhammad's (peace be upon him) noble character, molded by the teachings of the Quran, became a tradition and culture for Muslims. His honesty, courage, and humility earned him the title Al-Amin (The Trustworthy) even before his prophethood.

Allah says: *"And indeed, you are of a great moral character."* (Quran, 68:4)

"Just as We have sent among you a messenger from yourselves reciting to you Our verses and

purifying you and teaching you the Book and wisdom and teaching you that which you did not know." (Quran, 2:151)

---

2. Health and Diet: A Sunnah for All Generations:

The Prophet (peace be upon him) emphasized personal health and recommended practices for physical and spiritual well-being. His teachings on moderation in eating, the use of honey, and the consumption of dates are timeless examples of divine wisdom.

The Prophet's sayings (hadiths) highlight these principles:

"A human being fills no worse vessel than their stomach. The son of Adam needs only a few bites to keep him alive. If he must fill it, then one-third for his food, one-third for his drink, and one-third for his breath."

(Reported by Tirmidhi)

The Quran supports this wisdom:

"From their bellies (of bees) comes a drink of varying colors, in which there is healing for

people. Indeed, in that is a sign for a people who reflect."

(Quran, 16:69)

Modern science has validated the health benefits of honey, dates, and moderation, proving the timeless relevance of the Prophet's (peace be upon him) guidance.

II.   The Divine Favor of Allah

A. Toward the Disciples of the Prophets

1. Companions of Solomon (peace be upon him):

Solomon's companion, who brought the throne of Queen Bilqis almost instantaneously, demonstrated Allah's favor through extraordinary knowledge and power.

Allah Says:

"A person who had knowledge of the Scripture said, 'I will bring it to you before your glance returns to you.' And when Solomon saw it placed firmly before him, he said, 'This is from the favor

of my Lord to test me whether I will be grateful or ungrateful.' " (Quran, 27:40)

---

2. Companion of Muhammad (peace be upon him): Ali ibn Abi Talib:

Ali, a devoted follower and family member of the Prophet, exemplified Allah's favor through his proximity to the Messenger and his steadfast faith.

B. Toward Famous Kings

1. Dhul-Qarnayn:

Dhul-Qarnayn was granted immense power and knowledge by Allah to establish justice and benefit humanity.

"Indeed, We established him upon the earth, and We gave him a way to achieve everything."

(Quran, 18:84)

2. Pharaoh of Egypt:

Despite his tyranny, Pharaoh was used as a lesson for humanity. Allah preserved his body as a sign, demonstrating both His favor and His justice.

"So today We will save you in body that you may be to those who succeed you a sign."

(Quran, 10:92)

The Wisdom: Allah's favor is boundless, extending to both the righteous and the disbelievers.

- For the righteous, it is a means to achieve good and fulfill their duties.

- For the wicked, it serves as a reminder of Allah's ultimate authority and justice.

Conclusion: This chapter has explored the wisdom of Allah through His creation, the miracles of the prophets, and His divine favor. Whether through the accomplishments of

modern science or the lessons of history, Allah's signs are ever-present.

The next chapters will continue to unveil Allah's blessings and guidance, offering a deeper understanding of His mercy and omnipotence.

CHAPTER III:

Insights Into Allah's Favors
Upon 21st-Century Scientists

The 21st century's extraordinary scientific advancements are clear manifestations of Allah's boundless wisdom and favor upon humanity. These advancements, spanning fields such as astronomy, biology, physics, and technology, reveal the intricate design and mercy of Allah's creation. They not only benefit humanity in countless ways but also testify to the divine truths conveyed in the Quran over 1,400 years ago.

This chapter explores the harmony between modern discoveries and Quranic wisdom, inviting readers to reflect on how Allah's favors continue to guide and inspire humanity through science and innovation.

---

1. The Creation of Humans

The creation of humankind from a fetus, following Adam's creation from clay, reflects Allah's unmatched power and divine science. This creation process has always stimulated human curiosity, especially that of renowned scientists. Praise be to Allah for His infinite mercy upon us, His creations! The Quran says:

"Indeed, the example of Jesus to Allah is like that of Adam. He created him from dust; then He said to him, 'Be,' and he was. The truth is from your Lord, so do not be among the doubters." (Quran, 3: 59-60)

"And Allah created you from dust, then from a sperm-drop; then He made you mates. "

(Quran, 35:11)

Analysis: Let's reflect upon the wisdom and skill of Allah, our Creator, as demonstrated above in these two distinct methods of human creation. Why did Allah reveal this so clearly to His Prophet (peace be upon him) for us to understand? Surely, it was to invite humanity and scientists in particular to ponder over His creation, which is rooted in divine science.

There is no doubt that scientists have uncovered significant insights about human existence and maternal development. The sciences of chemistry and biology have explained the development of a child, from the sperm stage to birth. However, the Quran revealed these truths 14 centuries ago:

"We created man from an extract of clay; then We made him a drop of fluid in a secure resting place; then We made the drop into a clinging clot; And We created the clot into a lump And We created the lump into bones, and We clothed the bones with flesh; then We developed him into another creation. So blessed is Allah, the best of creators."

(Quran, 23:12-14)

Just as the creation of humankind demonstrates Allah's infinite wisdom, so too does the natural world, from the movements of the heavens to the intricate processes of life, all of which testify to His divine favor and mercy.

2. The Science of Astronomy

Modern astronomy reveals the vastness of the universe and the precise movements of celestial bodies—truths that the Quran conveyed centuries ago. Allah the Most High says:

"It is He who created the night and the day and the sun and the moon; all [heavenly bodies] in an orbit are swimming."

(Quran, 21:33)

This verse describes the celestial movements that modern astronomy has confirmed through advanced telescopes and satellite technology.

Furthermore, the Quran highlights the sun's specific trajectory:

"And the sun runs on its fixed course for a term [appointed]. That is the decree of the Almighty, the All-Knowing."

(Quran, 36:38)

Astronomers now understand that the sun moves through the Milky Way galaxy along a predetermined path—a fact revealed in the Quran long before the advent of modern science. Just as the celestial bodies follow precise orbits, reflecting Allah's divine order, other fields of science unveil similar intricacies, demonstrating the harmony in His creation.

---

3. Physics and the Structure of the Universe

Physics has equipped humanity with the means to understand the universe's structure, from the microscopic world of atoms to the vast expanse of galaxies. The Quran also emphasizes the balance and grandeur of Allah's creation:

"The creation of the heavens and the earth is greater than the creation of mankind, but most of the people do not know."

(Quran, 40:57)

This verse underscores the vastness of the cosmos, inviting reflection on the complexity and magnificence of Allah's creation. Just as physics unveils the intricacies of the universe's structure, the study of biology reveals the wonders of life, all of which testify to Allah's infinite wisdom.

4. Biology and the Mystery of Life

The Quran points to the origin and interdependence of life: *"And We made from water every living thing. Then will they not believe?"* (Quran, 21:30)

Modern biology confirms that water is fundamental to life on Earth. The study of DNA, cellular structures, and ecosystems reveals the intricate design of living organisms—a testament to Allah's infinite wisdom. Just as biology uncovers the mysteries of life, technology demonstrates humanity's ability to harness Allah's creation for progress and innovation, aligning with His divine plan.

5. Technology: A Modern Blessing

From communication tools to transportation, technology has transformed human life. These advancements align with Allah's divine plan and are foreshadowed in the Quran: *"And He created the horses, mules, and donkeys for you to ride and as adornment. And He creates that which you do not know."*

(Quran, 16:8)

The phrase "that which you do not know" anticipates modern inventions such as cars, airplanes, and other modes of transportation. Through the blessings of technology, humanity continues to witness Allah's mercy and wisdom, underscoring the importance of using these advancements responsibly as stewards of His creation.

Conclusion: Allah's blessings upon all His creations, reflected in science and technology, underscore His boundless mercy and infinite wisdom:

- Among scientists, there are those who do not believe in Allah. Yet, Allah does not exclude them from the blessings of this world or from the knowledge He grants. He allows them to conduct research and succeed in their enterprises.

- Believers in Allah, particularly Muslims, understand that all achievements are by Allah's will.

"Glory be to God, who taught mankind what he did not know."

(Quran, 96:5).

So they benefit from the advancements of Science such as:

1- Meteorological predictions of rain, snow, and wind.

2- Traveling by air to perform one of the five pillars of Islam—pilgrimage to Mecca.

- But one must marvel at the miraculous journey of Prophet Solomon (peace be upon him). God says:

"And to Solomon [We subjected] the wind—its morning [journey] was that of a month, and its evening [journey] was that of a month. "

(Quran, 34:12)

- Even more astonishing is the ascension of angels who ascend to the heavens and beyond to receive divine orders and revelations from their Lord.

"From Allah, the Lord of the ways of ascent. The angels and the Spirit will ascend to Him during a Day the extent of which is fifty thousand years."

(Quran, 70:3-4)

- All mankind, we must acknowledge science as originally stemming from God Almighty long before it became a human endeavor; and fulfil our role as a steward of his creation.

God says:

"And [mention, O Muhammad], when your Lord said to the angels, 'Indeed, I will make upon the earth a successive authority (khalifah).

(Quran 2:30)

Glory to God, the Most High and Exalted! And to His messengers His peace and blessings.

ANALYTICAL TABLES
ON HUMAN SCIENCE

Has Man Created or Merely Utilized the Materials of the Universe to Develop His Science?

This critical question examines the distinction between creation and utilization. Below is an analytical breakdown to provide clarity:

Aspect	Human Contribution	God's Role
Source of Materials	Man did not create the raw materials (iron, water, air, earth, etc.).	God created the universe and all it contains for human use (Quran, 45:13).
Discovery vs. Creation	Humans discover and utilize existing materials.	God is the Creator of all things (Quran, 20:6)
Knowledge and Skills	Humans acquire knowledge and develop skills	God endowed humans with intelligence and the ability to

Aspect	Human Contribution	God's Role
	through observation.	learn (Quran, 96:4-5).
Scientific Advancements	Humans develop technology (e.g., airplanes, medicine).	God allows these advancements through His will and decree (Quran, 2:255).
Ultimate Authority	Human science is limited and fallible.	God's knowledge is infinite and perfect (Quran, 6:59).

Conclusion: Man has not created anything ex nihilo (from nothing); rather, he has used what God has provided to develop his science. Every human achievement is a direct result of God's creation and permission. Therefore, scientific progress must be seen as a means to recognize God's greatness and to fulfill our duties of gratitude and worship. As the Quran states: *"Do they not look into the realm of the heavens and the earth and everything that God has created?"* (Quran, 7:185)

ANALYTICAL CHART
ON HUMAN SCIENCE

Compact Flowchart: God is the First and the Last

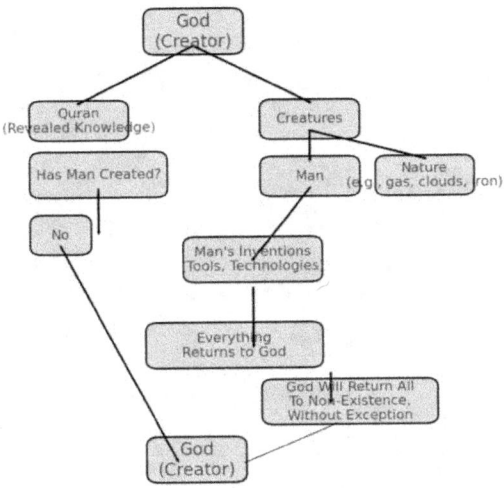

This flowchart illustrates how human achievements in science rely entirely on God's creation and permission.

Wisdom: God Almighty will return all to nonexistence without exception!

"Everyone upon it [the earth] will perish, and there will remain the Face of your Lord, Owner of Majesty and Honor." (Quran 55:26–27)

CHAPTER IV:

Reflections on the Significance of Science in Islam

The Compatibility of Science and Faith

In Islam, science and faith are not opposing forces. Rather, they complement one another, as both are rooted in the divine wisdom of Allah. The Quran encourages believers to observe, reflect, and study the world around them, emphasizing that such endeavors will lead to a deeper understanding of Allah's creation.

Allah the Most High says:

"We will show them Our signs in the horizons and within themselves until it becomes clear to them that this is the Truth. Is it not sufficient concerning your Lord that He is, over all things, a Witness?"

(Quran, 41: 53)

This verse highlights the link between scientific inquiry and spiritual awareness. As humanity uncovers the mysteries of the universe, the evidence of Allah's existence becomes even more apparent.

Science as an Act of Worship

Pursuing scientific knowledge is an act of worship in Islam, provided it is undertaken with the intention of benefiting humanity and strengthening one's faith. The Quran praises those who seek knowledge, describing them as people of understanding:

"Indeed, in the creation of the heavens and the earth and the alternation of the night and the

day are signs for those of understanding."
(Quran, 3:190)

The practical applications of science have greatly facilitated Islamic practices, bridging faith and technology. For example:

- **Adhan apps** allow Muslims worldwide to determine accurate prayer times.

- **Compass tools and Qibla direction** devices assist in orienting toward Mecca.

- **Online platforms and financial systems** have simplified Zakat donations, enabling Muslims to fulfill their obligations more efficiently.

The Quran's emphasis on reflection and innovation makes these advancements a natural outcome of divine guidance. The Prophet Muhammad (peace be upon him) also emphasized the importance of learning, stating:

"Seeking knowledge is an obligation upon every Muslim." (Sunan Ibn Majah)

This call to seek knowledge has inspired generations of Muslim scholars, whose contributions to science, mathematics, medicine, and philosophy laid the foundation for much of the modern world's intellectual progress.

Misconceptions about Science and Religion

In some circles, there is a misconception that science and religion are incompatible. However, Islam provides a framework that integrates scientific exploration with spiritual values. The Quran not only encourages the study of natural phenomena but also provides insights that align with scientific discoveries.

For instance, the Quran describes embryonic development with remarkable accuracy:

The Quran states:

"We created man from an extract of clay. Then We made him a drop in a secure place; then We made the drop into a clinging clot, and We made the clot into a lump, and We made [from] the lump bones, and We covered the bones with

flesh; then We developed him into another creation. So blessed is Allah, the best of creators."

(Quran: *12–14)*

This description, revealed over 1,400 years ago, corresponds with modern scientific findings about human development.

Wisdom:

The integration of science and faith is a hallmark of Islam. By pursuing knowledge with the right intentions, Muslims can strengthen their connection to Allah, contribute to the advancement of society, and fulfill their responsibilities as stewards of the earth.

This chapter serves as a reminder that the pursuit of science, when guided by faith, is a noble endeavor that aligns with the principles of Islam.

The DOUBT:

Eliminating Doubt about the Acceptance of Science within Islam

Over the centuries, some Muslim communities have harbored skepticism or disdain toward science and scientists. Even today, some Muslims hesitate to accept scientific advancements from a religious perspective. They avoid discussing or engaging with topics that bridge the Quran and science. One of the reasons—if not the only one— is that, until a certain era, the majority of scientists did not believe in Allah, the One God and Creator of the universe.

However, we all know that in this world, no one is excluded from the blessings and happiness distributed by Allah. We also recognize that diversity and tolerance are key to the development of science, which knows no boundaries

---

Analysis

Dear Muslim brothers and sisters, we all understand that the prophets (peace be upon them all) practiced science through miracles to fulfill their missions—always with the permission of Allah, the Most High. Notably, the last of them, Prophet Muhammad (peace be upon him), delivered information on the majority of essential elements of the universe through the Quran, offering key insights that even today's scientists are gradually uncovering.

"It is He who has created the heavens and the earth in truth, and the day He says, 'Be,' it will be. His word is the truth."

(Quran 6: 73)

This understanding should inspire Muslims to embrace the compatibility of science and Islam, reinforcing faith while exploring Allah's creation.

Addressing Doubt through Quranic Verses

Science and faith are often perceived to be at odds, yet the Quran encourages reflection, inquiry, and learning.

Misunderstandings about this relationship often lead to doubt among believers. However, by exploring Quranic verses and Islamic history, it becomes clear that science not only aligns with faith but also serves as a powerful tool to strengthen it.

1. Doubt and Reflection

The Quran explicitly invites believers to reflect upon themselves and their surroundings, demonstrating that inquiry is an integral part of faith.

Allah the Most High says:

"Do they not reflect upon themselves? Allah created the heavens and the earth and

everything between them in truth and for an appointed term. Yet many among the people deny they will meet their Lord."

(Quran, 30:8)

This verse encourages believers to explore their existence and the natural world, framing scientific study as an act of worship that deepens one's understanding of Allah's creation.

2. The Wisdom of Accepting Science

The fear that science may contradict faith often stems from misunderstanding. However, the Quran itself is rich with knowledge and references to natural phenomena, urging believers to study them.

Allah says:

"We will show them Our signs in the horizons and within themselves until it becomes clear to them that this [Quran] is the truth; Is it not sufficient concerning your Lord that He is, over all things, a Witness? "

(Quran, 41:53)

The wisdom behind these verses serves as a guiding principle for integrating science and faith.

It encourages believers to embrace scientific exploration as a means of recognizing the signs of Allah in the universe. Rather than fearing a contradiction between science and religion, these verses inspire confidence in the harmony between the two, affirming that all discoveries ultimately point back to God Almighty, the Creator of all. By acknowledging the powerful hand of God in every aspect of creation, believers can strengthen their faith while contributing to the advancement of knowledge and understanding.

---

3. Overcoming Misconceptions

Some Muslims mistakenly view science as contradictory to Islam. This perception arises from a lack of understanding of both the Quran and scientific findings. By studying both, these apparent contradictions dissolve.

Even non-Muslim scientists, despite not recognizing their discoveries as evidence of Allah, often validate Quranic truths through their work: Allah says:

"But Allah favors whom He wills among His servants." (Quran, 14:11)

This verse reminds us that intelligence and creativity are divine gifts distributed among all humanity, regardless of faith. Muslims can learn from and expand upon these contributions.

The Role of Scientists in Clarifying Doubt

1. Contributions of Muslim Scientists

During Islam's Golden Age, scholars like Al-Khwarizmi, Ibn Sina, and Al-Biruni exemplified the harmony between faith and science. Their groundbreaking contributions to mathematics, medicine, and astronomy not only advanced human knowledge but also showcased how science can strengthen Islamic teachings.

2. Quranic Alignment with Modern Science

Numerous Quranic verses align with modern scientific discoveries, illustrating the divine wisdom within the scripture.

For instance, the Quran states:

"And We made the sky a protected ceiling, but they, from its signs, are turning away.

" (Quran, 21:32)

This verse refers to the protective nature of Earth's atmosphere, a concept only understood through modern scientific exploration

---

Wisdom: The doubt surrounding the compatibility of science and Islam is rooted in misunderstanding, not contradiction. The Quran not only encourages exploration and reflection but also aligns with scientific truths, proving its divine origin. By embracing both faith and science, Muslims can overcome doubt, honor their rich intellectual heritage, and contribute meaningfully to humanity's advancement.

CONCLUSION

What is fundamental in this life is sincere belief in God, His holy books, and His prophets—starting from Adam, Abraham, Ishmael, Moses, Jesus, Solomon, and Muhammad (peace be upon them all). The Quran affirms:

"The Messenger has believed in what was revealed to him from his Lord, and so have the believers. All of them have believed in Allah, His angels, His books, and His messengers, saying, 'We make no distinction between any of His messengers.' " (Quran.2: 285)

Science, in its essence, is a sacred favor from God and a test for us humans, enabling us to find the right path leading to Him.

It is essential to understand that God's will is always dominant. The Quran serves as a straightforward guide with significant signs.

Nowadays, it is accessible to everyone, including scientists, through whose hands some of God's signs have been realized or explained—such as:

the separation of the two waters (of the river and the sea) that do not mix, the movement of our galaxy, and the certainty of a child's development in the womb of its mother. The Quran states:

"We will show them Our signs in the universe and within themselves until it becomes evident

to them that this is the Truth."

(Quran.41:53)

Acknowledgments

I extend my gratitude to all readers, especially young Muslims. And

May the Benevolent God grant us the spirit of reflection and analysis—a scientific mindset that brings us enlightenment and strong faith in God and His divinity, just as those Muslims who are described in the following verse: *"Who remember Allah while standing, sitting, and lying on their sides and reflect on the creation of the heavens and the earth, saying: 'Our Lord, You did not create this aimlessly; glory be to You! Protect us from the punishment of the Fire."*

(Quran 3: 191)

God is Great and Most Exalted!

Sincere prayers upon the honorable Prophet Muhammad (peace and blessings be upon him).

Quran's Wisdom and Modern Science

Solar System

"And He subjected the Sun and the Moon all swimming in an Orbit."

بسم الله الرحمن الرحيم

Earth and Life

" And He taught Men what he didn't know."

"And your Lord inspired the bee."

The Quran states:
" Indeed, in the creation
 of the heavens and the earth, and the alternation of the
night and the day, and the [great] ships which sail through
the sea with that which benefits people, and what Allāh has
sent down from the heavens of rain, giving life thereby to the
earth after its lifelessness and dispersing therein every [kind
of] moving creature, and [His] directing of the winds and the
clouds controlled between the heaven and earth are signs for
a people who use reason." (Quran 2:164)